THE AUDUBON SOCIETY POCKET GUIDES

A Chanticleer Press Edition

Ann H. Whitman
Editor

Kenn Kaufman
John Farrand, Jr.
Consultants

Western Region

FAMILIAR BIRDS OF NORTH AMERICA

Alfred A. Knopf, New York

Prepared and produced by Chanticleer Press, Inc.,
New York.
Color reproductions by Nievergelt Repro AG,
Zurich, Switzerland.
Typeset by Dix Type Inc., Syracuse, New York.
Printed and bound by Dai Nippon, Tokyo, Japan.

First Printing.

Library of Congress Catalog Number: 86-045589
ISBN 0-394-74842-5

Trademark "Audubon Society" used by publisher under
license from the National Audubon Society, Inc.

Cover photograph: Western Tanager by Betty Randall

Contents

How to Use This Guide

Of all wild creatures, birds are the most conspicuous and colorful. Even in our crowded cities, there are many kinds of birds, easily noticed and identified. Learning to identify birds is an enjoyable pursuit—an end in itself or the first step in an intensely absorbing hobby that can last a lifetime.

Coverage

This new guide covers 80 of the most frequently encountered and abundant birds of the West. Our range is bounded by the Pacific Ocean on the west and on the east by the 100th meridian, extending from Edwards Plateau in Texas northward through Oklahoma and westward through eastern Colorado, Wyoming, and Montana, and in Canada, along the eastern foothills of the Rocky Mountains. Thus, the boundary roughly follows the Rockies. The companion volume to eastern birds covers species east of this line.

Organization

This easy-to-use guide is divided into three parts: introductory essays; the color plates and species accounts; and appendices.

Introduction

As a basic introduction, the essay "Identifying Birds" outlines the field marks you should notice when you look at a bird. "Bird-Watching" offers expert advice on when, where, and how to look for birds. Finally, "Attracting Birds" provides tips on how to make your backyard a haven for colorful songsters.

The Birds

This section contains 80 color plates arranged visually, by the color, shape, and overall appearance of each species. Here are a selection of common water birds, shorebirds, raptors, game birds, and songbirds. Facing each illustration is a description of the species' important field marks as well as information about its voice, habitat, and range. A map showing nesting and winter ranges supplements the range statement. The introductory paragraph discusses other information, such as behavior, nesting habits, and close relatives.

Appendices

Featured is an essay on the 36 families of birds represented in this book (out of a total of 69 North American families). Knowing family traits helps to identify quickly birds covered in this guide as well as other close relatives. The last feature is an illustrated glossary defining common terms that may be unfamiliar.

Whether your home is in a remote countryside or in a populous city, you will derive hours of pleasure observing and learning about birds.

Identifying Birds

When you see a new and unfamiliar bird, what points should you notice to help you in naming it? By going from important general features to specific ones, you will be able to narrow the identification down to a few clear choices.

Size
Exact size in inches is very difficult to judge in the field. You can usually get a general impression, however, if you compare the size of a new bird to that of something familiar, such as a House Sparrow, a robin, a crow, or some larger bird.

Shape
Experts can name most birds by shape alone. Even for beginners, learning to analyze shape helps to place a bird in the correct group.
Overall, is the bird slim or chunky? Are its neck and legs long or short? Is its tail short, medium, or long? Is it forked at the tip, square-ended, rounded, or pointed? Bill shape is one of the most important features in placing a bird in the correct family: Is the bill short or long? Thick or thin? Straight or curved? When the bird flies, are its wings long and pointed, short and rounded, or some other shape? What is the shape of its head? Does it have a crest?

8

Behavior

What a bird does often helps reveal what it is. Try to notice the way a bird perches, how it forages for food, what it eats, how it flies, and so on.

Color and Pattern

Try to look at the bird systematically, noting the crown, face, throat, underparts, wings, tail, and upperparts. Is the crown broadly striped, finely streaked, or plain? Is there a central spot of color, a pale ring around the eye, a stripe over the eye, or a dark ear patch? Does the throat contrast with the face, or with the breast? Is it framed by dark "mustache" stripes?
Are the underparts streaked (lengthwise), barred (crosswise), spotted, or plain? Is there any contrasting color under the base of the tail?
Are there contrasting wing bars? How many, and what color? Are there contrasting outer tail feathers, a band at the tip, or spots at the corners? Is the back streaked or plain? Is there a pale patch on the rump (just above the tail)?

Voice

In some difficult groups, voice is the best way to separate similar species. And for most birds, it is useful to make a notation of any songs or calls that you hear.

Bird-watching

It is possible to see birds practically anytime and anywhere, but certain times of day, places, and seasons are undeniably better. Knowing when to look and how to search can make your bird-watching more exciting. There is no need to travel far; wherever you live, there are undoubtedly some good birding areas nearby. Local birders may direct you to favorite spots, but even without such advice you will soon recognize good places.

Habitat

All birds need food, water, and shelter, but no two species have precisely the same needs. Thus variety in the habitat is the key to finding a variety of birds. For example, a mixed forest is usually better than a stand of just one kind of tree; a forest with trees of all ages, brushy undergrowth, clearings, and dead timber will be even better. The edge of a forest—or any place where one habitat meets another—is often a spot where birds are most numerous and easiest to see.

Water improves a good birding area, not only for aquatic species but also for land birds, which may come to drink. Here again, variety is important: Ponds or rivers with heavy vegetation, marshy inlets, and muddy estuaries are always promising.

Birds also gather where food is locally abundant, and

concentrations of fruiting or flowering plants may attract many different species. Garbage dumps are often excellent places to study gulls and other scavengers. A few species are adapted to simple environments, such as short-grass fields or the open ocean. Visit as many habitats as possible, concentrating on those that are richest in variety but not neglecting the others.

Time

When you go birding, consider the time of day. Most land birds are active very early in the morning and again in the evening, but they can be very quiet at midday. In cold or wet weather, small birds tend to remain active all day. Water birds have schedules that may be governed more by the tides than by time. Many rest at high tide and feed when the water rises or falls.

Seasons

Each season of the year has its own advantages for bird-watching. Winter can be a good time to begin, because—especially in the north—fewer species are present, allowing one to learn the birds gradually. Wintering birds tend to form flocks at good feeding areas, such as hedgerows and orchards. Spring is the favorite season for many bird-watchers. In the East, most spring migration occurs between early April and late May; the northbound migrants are in their most colorful plumage, and many sing on their way. Summer is the breeding season for most species, and you can watch birds build their nests and raise their young. The fall migration lasts much longer than that of spring; the first southbound shorebirds appear by early July, and the last of the waterfowl are still moving south in early December. There are more birds then, because populations are swelled by the young hatched in summer. Autumn is challenging, because many species are harder to identify in fall plumage.

Birding Techniques   Try to walk slowly and smoothly, because birds are likely to be alarmed by sudden motions. Learn to watch vegetation for movements that reveal the feeding actions of small birds.

Using Binoculars   The best way to see field marks is to use binoculars, but many beginners have trouble with them. A good technique is to stare fixedly at a bird while raising the binoculars into your line of sight; this will aim them correctly. Mastering this method will make your bird-watching more rewarding.

In choosing binoculars, look at the specifications—for example, 7×35. The first number is the magnification; in this example, the image will be enlarged seven times.

12

The second number is the diameter of the objective (forward) lenses, which governs the amount of light allowed in. Higher-powered binoculars need more light, and generally this second number should be at least five times the first. This rule is less important in very high-quality (and high-priced) binoculars. You should also consider the minimum focusing distance: To look at a small bird just a few yards away, you will need to have binoculars that allow you to focus that close.

Calls and Songs

Birds utter two basic kinds of vocalizations—songs and calls. Songs, usually complex, are used mainly by adult males during the breeding season to establish territories or attract mates. Calls are usually simple notes, single or repeated, given at all seasons to express alarm or to maintain contact. All songs and most calls are distinctive; concentration and practice are the keys to recognizing them. You can often fix a bird's voice in your memory by describing it to yourself, using the transcriptions in this book as a guide. Practice calling birds to you: Squeaking or "shushing" noises often arouse curiosity. Above all, listen. Experienced observers find and identify many of their birds by sound.

Attracting Birds

Knowing what birds need is helpful when you go looking for them, and you can apply this knowledge to bring birds to your garden. By providing food, water, and shelter, you can easily attract a lively assortment of birds year after year.

Feeders
Bird-feeders range from the simple to the elaborate. One common arrangement is an open seed tray with tiny drain holes in the bottom, and perhaps a small roof to keep off rain and snow. Seed feeders attract thick-billed birds like finches and sparrows, as well as chickadees, nuthatches, and others. Provide both small and large seeds; thistle, millet, and sunflower seeds make an ideal combination, or you can add cracked corn or raisins for variety. A piece of fruit on a spike beside the feeder may bring in robins and mockingbirds.

Some people like to put out suet, which should be enclosed in wire mesh to keep large birds or squirrels from making off with the whole piece. Suet feeders attract tree-climbers, like woodpeckers and nuthatches; many birds find peanut butter an acceptable substitute. Hummingbird feeders can liven up any garden in summer, and may be used all year in warmer regions. These feeders—red tubes of plastic or glass, filled with sugar-water—are sold at garden-supply stores. Regular maintenance and cleaning are important.

Squirrels
At feeders, squirrels can be pests, but they may often be foiled by strategic planning. Place the seed tray atop a smooth pole and attach a broad cone or disk of sheet metal to the pole some distance below it. Some hanging feeders are also safe.

Baths
Birds like to visit a garden bath. A bird-bath should be very shallow with a sloping bottom, and should be cleaned regularly. Resident birds will quickly find a bird-bath, but to attract transient birds it helps to rig up a small fountain so that the sound of water is audible.

Shelter
Ultimately, the best way to bring birds to your yard is to plant the trees and bushes that provide natural shelter. Evergreens, thick bushes, and vines provide a year-round haven; mulberry, multiflora rose, and other plants that also furnish berries are doubly attractive to many species. Put your feeder near enough to these plants that the birds can easily flee from danger, but not so close that cats or other predators can ambush unwary birds. And if you allow a quiet corner to grow up in annual weeds, it will be perennially popular with the birds.

THE BIRDS

Great Blue Heron *Ardea herodias*

In the shallow waters of a marsh or pond, this magnificent bird is often seen standing motionless, waiting patiently for the movement of a fish, frog, or salamander to break the stillness of the water's surface: Then it strikes with its long, pointed bill. The Great Blue Heron builds a flimsy platform nest of sticks; its colonies may include just a few pairs or several hundred birds. It usually dwells in the tops of tall trees; in California, it has been known to nest 100 feet up in tall redwoods.

Identification   39–52″. Large, with wingspan up to 7′. Grayish, with long yellowish bill; head white, with black on crown and nape; legs dark. Immature similar to adult but paler, with all-black crown.

Voice   A harsh croak; usually silent.

Habitat   Marshes, ponds, lakes, and rivers.

Range   Breeds throughout United States and S. Canada, but absent from deserts and high mountains. Winters throughout most of S. United States and along coasts north to S. Alaska and New England.

18

Common Loon *Gavia immer*

Built for an aquatic life, loons are clumsy on land. Their legs are set very far back on the body, an adaptation for diving that makes walking an awkward business. The Common Loon is famous for its demonic laughing call, which resounds through still summer nights in the North. These birds often swim long distances underwater; a loon may make a quick, silent dive in one part of a lake and pop up again 200 yards away.

Identification 28–35". Large. Breeding bird has black-and-white checkerboard pattern above, with velvety black head, white necklace, white underparts, and straight, black, daggerlike bill. Winter bird and immature blackish above, whitish below, with paler bill.

Voice Loud yodeling call on breeding grounds; wild, laughing call given at night.

Habitat Lakes and bays; farther out to sea in migration.

Range Breeds from Alaska to S. British Columbia and east to Newfoundland and New England; winters on Great Lakes and along coasts.

Western Grebe *Aechmophorus occidentalis*

Because of their skill in the water and inelegance on land, grebes were long thought to be related to loons. Despite their similar life-styles, however, the families are not related. In spring, Western Grebes perform an elaborate courtship display. Using their lobed feet like paddle wheels, the male and female race furiously, side by side, across the surface of the water, with their bodies nearly erect and their bills pointed skyward. A closely related species, Clark's Grebe (*A. clarkii*), has white on the face extending above the eye and a yellow bill.

Identification 20–24". Dark gray to blackish above, white below, with long, swanlike neck and black crown tuft; eyes red; bill long, straight, sharply pointed; dull greenish yellow.

Voice A loud, rolling *crreee-crreee.*

Habitat Large lakes and inland marshes; coastal waters in winter.

Range Breeds from S. British Columbia and N. Alberta to S. California and South Dakota. Winters along coast from N. British Columbia to Mexico.

Double-crested Cormorant *Phalacrocorax auritus*

The most abundant cormorant in North America, the Double-crested may gather in flocks of up to 2000 individuals. A heavy-set water bird, it sometimes plunge-dives for fish and other prey, but more commonly performs a surface dive and swims about underwater, coming up with its catch in its bill. Unlike most other water birds, cormorants lack waterproof plumage; to dry their feathers, they perch on a rock or piling and spread their wings wide in the sun.

Identification 30–36". Adult black with bare patch of orange skin beneath bill; breeding adult has small tufts on each side of crown. Immature brownish, paler on breast. Bill long, straight, with hooked tip in all plumages.

Voice Usually silent; gives grunting calls near nesting colony.

Habitat Lakes, rivers, and seashores.

Range Breeds along coast from Alaska to S. California, and inland in Saskatchewan, W. Montana, S. Idaho, and N. Utah. Winters mainly along coast. Also in the East.

Green-winged Teal *Anas crecca*

Although primarily a bird of marshes and mud flats, the Green-winged Teal is at home on land; it has even been known to visit upland woods to eat berries, grapes, and acorns. Despite its small size, this species is swift in flight. Green-winged Teals migrate early in spring; from their wintering grounds, they travel north in compact flocks, arriving in northern Alaska by early May.

Identification 12–16". Small. Breeding drake has chestnut-colored head with bright green patch extending back behind eye; vertical white stripe on side of breast; belly pale whitish; sides gray, wings darker gray. Female, immature, and nonbreeding drake mottled gray and brown, with whitish belly and dark upper wings. All have green speculum.

Voice Drake gives a clear, repeated whistle; female quacks.

Habitat Lake borders, ponds, marshes, and mud flats.

Range Breeds from N. Alaska south to N. California and Colorado, east to Newfoundland and N. New England. Winters from British Columbia and Utah south, east to Texas; also along Atlantic and Gulf coasts.

26

Mallard *Anas platyrhynchos*

Probably the most common duck in all of the Northern Hemisphere, the Mallard is the ancestor of nearly all the domestic ducks in the world. Even wild populations seem to find the company of people unobjectionable, for Mallards are frequently seen in harbors and large city parks, where semidomesticated family groups often paddle up alongside a boat, quacking and looking for a handout.

Identification 18–27". Breeding drake grayish, with green head, brown breast, and white neck-ring. Female and nonbreeding drake mottled brown with white tail and orange-and-brown bill. Both sexes have glossy blue speculum, visible in flight.

Voice Female utters a loud, familiar quack; male gives a reedy *rah-rah-rah.*

Habitat Ponds, lakes, rivers, marshes, bays, beaches, and parks.

Range Breeds from Alaska to Nova Scotia, south to S. California, Texas, and Virginia. Winters from British Columbia and New England to Mexico and south.

28

Cinnamon Teal *Anas cyanoptera*

Larger than the Green-winged Teal, the Cinnamon Teal inhabits marshy wetlands in dry regions, and is especially numerous near the Great Salt Lake. This duck often travels in flocks with the related Blue-winged Teal (*A. discors*), whose range overlaps the Cinnamon's in parts of the Northwest. The females of the two species are very similar, but the Blue-winged's breeding drakes can be distinguished by the white crescent on the face.

Identification 14½–17″. Breeding drake bright rufous with pale blue patch on each wing (may look whitish). Immature, female, and nonbreeding drake mottled brown with pale blue wing patch. Bill long, somewhat spoonlike at tip.

Voice Female gives a soft, high quack; male a low whistle; also clucking and chattering notes.

Habitat Marshes, prairie potholes, slow-moving streams with reedy vegetation, and other wetland areas.

Range Breeds from S. British Columbia to Mexico, and east to the Dakotas, Kansas, and Texas. Winters from California (Sacramento Valley) and S. Texas to N. South America.

30

American Coot *Fulica americana*

In warm weather, this chickenlike bird inhabits freshwater marshes and ponds, but with the arrival of freezing temperatures it makes its way to the seacoast in search of ice-free water. Unlike other members of the rail family, the American Coot has lobed feet that help make it an accomplished swimmer and diver. It usually associates with ducks.

Identification 13–16". Adult slate-gray with black head and neck setting off pale, ivory-colored bill and frontal shield (shield has red spot, conspicuous at close range). Immature similar to adult but paler, with dull bill.

Voice A variety of cackles, clucks, and low croaking notes.

Habitat Open freshwater marshes, ponds, and lakes; salt water in winter.

Range Breeds from British Columbia and N. Alberta to S. Ontario and New England, south to Mexico and Florida. Winters along coast from British Columbia to Mexico, east through southern states to Texas and Florida; also along Atlantic Coast.

Pied-billed Grebe *Podilymbus podiceps*

This cousin of the Western Grebe is only about half the size of that species. The most widely distributed grebe in North America, the Pied-billed is shyer than many of its relatives. When frightened, it submerges slowly beneath the water's surface, expelling air from both its feathers and lungs to help it to sink. It is the only grebe that does not show a white wing patch in flight.

Identification   12–15". Stocky, compact; brownish, with conspicuously short, chickenlike bill. Breeding bird has black throat and black ring around bill. Juvenile striped, with recognizable grebe bill.

Voice   A low, hollow series of *cow-oo, cow-oo* notes.

Habitat   Lakes, ponds, and marshes; open water in migration.

Range   Breeds throughout most of North America south of central Canada. Winters from British Columbia south to S. California and east to Texas and Florida; also along Atlantic Coast.

Canada Goose *Branta canadensis*

The Canada Goose is often heard before it is seen; in migration, long, wavy, V-shaped flocks give their musical honk from far overhead. In fall, these birds often descend on harvested fields, where they glean food from among the stubbly remains of the crop. The Canada Goose has a strong allegiance to its breeding grounds and returns year after year to nest in the same place; thus several different races have developed, differing chiefly in size.

Identification Large races 32–48″; small races 22–27″. Brownish-gray body and wings with long black neck and black head; distinctive white cheek patch usually extends under throat. Belly and flanks white. Bill and legs black.

Voice Larger races give a resounding *ornh-whonk*; smaller races cackle a high *ahnk*.

Habitat Lakes, rivers, marshes, bays, cornfields, and grasslands.

Range Breeds from N. Alaska to Baffin Island, south to NE. California, Missouri, and Ohio. Winters from southern part of breeding range to Mexico and Gulf Coast.

Tundra Swan *Cygnus columbianus*

This smallest North American swan is a common sight in winter along the Pacific Coast. Formerly known as the Whistling Swan, this species breeds in the remote tundra areas of northern Alaska and arctic Canada, where it builds a moundlike nest of mosses. Lewis and Clark discovered this species in the course of their famous expedition; it is named for the Columbia River, where they first encountered it.

Identification 47–58". Large, white, with neck held straight up. Bill and feet black; bill usually has small yellow spot at base near eyes. Cygnet tinged with light gray-brown, has pinkish bill.

Voice A high, bugling *wow-how-wow*, more musical than honk of Canada Goose.

Habitat Tundra; marshes; bays and seashore areas in migration.

Range Breeds from W. and N. Alaska to Baffin Island. Winters along coast from S. Alaska to S. California, and in parts of the Southwest; also on Atlantic Coast.

Brown Pelican *Pelecanus occidentalis*

The large Brown Pelican is highly gregarious, living in flocks most of the year; it flies with slow, powerful wingbeats in single file or a V formation. This species eats fish, which it scoops up in its large throat pouch, often plunge-diving for its prey from the air. High levels of DDT contamination in fish resulted in a drastic decline in the breeding populations of these birds; banning of DDT and other hard pesticides has helped the Brown Pelican to make a comeback.

Identification 45–54". Adults stocky and dark brown, with massive bill and throat pouch, and whitish head. Breeding birds have cinnamon-brown on back of neck. Young have dull brown heads.

Voice Adults silent; nestlings squeal and grunt.

Habitat Bays, beaches, estuaries, and lagoons.

Range . Breeds along California coast from San Francisco south; occasionally wanders northward. Also on Atlantic and Gulf coasts.

Western Gull *Larus occidentalis*

Except for the closely related Yellow-footed Gull (*L. livens*), the Western Gull is the only "sea gull" on the West Coast with dark wings and a dark back, and in many areas along the Pacific it is the only nesting gull. Like most members of its family, the Western Gull knows how to fish, but is often content to scavenge along beaches and visit garbage dumps, or to trail after fishing boats looking for scraps tossed overboard. It breeds in colonies on rocky offshore islands.

Identification 20–23". Adult dark gray above with black wing tips; clear white below, with white head and tail. Bill thick, yellow, with red spot on lower half toward tip. Legs pink. Immature varyingly mottled brown and gray; reaches adult plumage in its fourth winter.

Voice A series of raucous calls and squeals.

Habitat Beaches, islands, inlets, and harbors.

Range Along coast from Washington to Mexico; some birds move north to British Columbia in winter.

42

Sanderling *Calidris alba*

The Sanderling differs from other members of its family in that it lacks a hind toe; a look at the footprints can be a help in identification. During most of the year, small parties of these familiar little sandpipers can be seen running up and down a beach, searching for small mollusks and crustaceans exposed by the backwash of retreating waves. The Sanderling is a worldwide bird, nesting throughout the Arctic and migrating to southernmost Africa and South America.

Identification 7–8½". Small. Breeding adults have rufous head, back, wings, and upper breast, with blackish mottling; belly white. Winter plumage pale gray above, white below. Bill and legs black. Bold white wing stripe in flight.

Voice Flight call a sharp, high-pitched *kip* or *kit*.

Habitat Primarily along sandy beaches; also on tidal flats, lakeshores, and sandbars.

Range Breeds from Arctic Alaska to Baffin Island. Winters along coast from British Columbia to Mexico and south. Also along Atlantic Coast.

Spotted Sandpiper *Actitis macularia*

Perhaps the most familiar shorebird in North America, the Spotted Sandpiper can be found in almost any wet place, from beaches to ponds, streams, and rain pools. A good walker, this species is sometimes seen clambering over rocks and logs; it can also swim, and may even dive beneath the water's surface to escape an approaching hawk. In coastal areas, its diet is suitably marine, but inland it takes a variety of beetles, grasshoppers, and other insects, as well as small, young freshwater fish.

Identification 7–8". Small, trim; gray-brown to olive-brown above; breeding adults boldly spotted below. Winter birds and juveniles white below with brown smudge on neck and breast. White wing stripe conspicuous in flight.

Voice A clear *peet-weet* or *weet-weet*; also a soft trill.

Habitat Anywhere with water: beaches, bays, wet meadows, streams, lakes, and ponds.

Range Breeds from Alaska to S. California, east throughout most of North America. Winters along Pacific Coast from British Columbia south; also in the Southeast.

46

Killdeer *Charadrius vociferus*

Comfortably adapted to life in almost every part of North America, the Killdeer is often heard crying out its name, *kill-deer, kill-deer*, in a long series of constantly repeated phrases. Killdeers make a nest in a shallow depression in the ground. They often nest in pastures, where chicks are vulnerable to the heavy footfalls of wandering livestock; to protect its young, a Killdeer may fly up suddenly into the face of an intruding cow.

Identification 9–11″. Robin-size. Grayish brown above, white below, with 2 distinct black breast bands; rufous rump and tail conspicuous in flight. Bill thin, black.

Voice A shrill, ringing *kill-deeeer, kill-deeeer*, repeated continuously; also a shorter *dee, dee, dee.*

Habitat Fields, pastures, mud flats, beaches, and other open areas near water.

Range Breeds from central Alaska and Newfoundland south to Mexico and Florida. Northern populations move to southern part of breeding range in winter; southern birds nonmigratory.

48

Common Snipe *Gallinago gallinago*

A retiring bird of marshes and other wetlands, the Common Snipe feeds chiefly on insects; it probes in soft mud with its long, sensitive bill, and can lift the upper mandible to grasp its food firmly. The Common Snipe performs an impressive display dive from 300 feet in the air; as it plunges earthward, its outer tail feathers vibrate and produce a hollow "winnowing" sound that sometimes can be heard for half a mile.

Identification 10½–11½". Long-billed, slender, brownish shore bird; mottled upperparts and head have prominent buffy stripes; belly white, breast and flanks with heavy bars and spots. In flight, has long, pointed, dark wings.

Voice A sharp, grating *scaip* when flushed; a high *wheet-wheet* on breeding grounds.

Habitat Marshes, ponds, bogs; occasionally in salt marshes.

Range Breeds from Alaska east to Labrador, south to central California and Massachusetts; sometimes father south. Winters from southern part of breeding range to South America.

50

Curve-billed Thrasher *Toxostoma curvirostre*

A resident of desert brushland in the Southwest, the Curve-billed Thrasher is often heard whistling from a perch or in flight; its distinctive call has an almost human quality. Like most other members of the mimic-thrush family, this species feeds on the ground; it uses its long bill to toss leaf litter aside as it searches for insects.

Identification 9½–11½″. Grayish brown above with long tail; faintly spotted below. Bill long and downcurved; eyes yellow or orange. Juvenile similar but has shorter, straighter bill, dark eyes, and even paler spots below.

Voice Call a whistled, loud *whit-wheet?*

Habitat Brushy desert areas, especially with cactus and thorn scrub; also brushy desert streamsides and residential areas.

Range Breeds from central Arizona east to Texas and south to Mexico. Winters from southern part of breeding range southward.

Northern Mockingbird *Mimus polyglottos*

A common bird in open countryside, the Northern Mockingbird sings throughout the year, seeming never to tire of performing its extensive repertoire of imitations and musical phrases. This species' skillful mimicry is not restricted to the songs of other birds; the Mockingbird has been known to imitate the noises of farm machinery, clucking hens, and even the sound of a piano. The Mockingbird is very territorial, and often chases other species of birds away from promising sources of food.

Identification 11". Soft gray above, paler gray to whitish below, with white wing bars and long tail. In flight, wings show bold white patches; tail shows white borders.

Voice Song a rich, infinitely varied medley of harsh and musical notes, interspersed with imitations and repetitions.

Habitat Woodland edges, desert scrub, cities, farms, suburbs.

Range Breeds from N. California to S. Wisconsin and Nova Scotia, south to Mexico and Florida. Northernmost populations move southward in winter.

Western Kingbird *Tyrannus verticalis*

This insect-feeder is abundant in most of the West, especially on rangeland and in agricultural areas, where it perches atop trees and poles. Like other members of the tyrant-flycatcher family, it is not afraid to attack larger birds, such as crows and hawks, when these intruders unwittingly approach the nest. The Western Kingbird catches its food on the wing, darting swiftly out from its perch to snatch a flying insect.

Identification 8–9½". Head, neck, and upper breast pale gray (darkest on crown); back darker gray with greenish wash; lower breast and belly yellow. Tail long, black, with white outer edges. Red feathers on crown usually concealed.

Voice A loud *whit* or *kit*; also various shrill calls and chattering notes.

Habitat Chaparral, orchards, pinyon-juniper woodlands, and other open areas with scattered trees.

Range Breeds from S. British Columbia and Manitoba to S. California and SW. Texas; absent from Pacific Northwest. Winters in Central America.

Purple Martin *Progne subis*

In the West, this large swallow shuns man-made martin houses, seeming to prefer tree cavities in the open countryside, or suitable spots in cities. In parts of the desert Southwest, it occupies abandoned woodpecker holes in the trunks of saguaro cactus. Purple Martins sometimes fly low over a river or pond, skimming the surface to catch a drink and a quick bath on the wing.

Identification 7–8¼". Adult male glossy blue-black overall. Female and immature slightly duller above and pale gray below. Tail in all plumages long and shallowly forked.

Voice Song a series of rich, gurgled notes; also a clear *tee-tee-tee*.

Habitat Open woodlands, farm country, and residential areas.

Range Breeds along Pacific Coast from S. British Columbia to Mexico; also from E. British Columbia and Manitoba east through Great Lakes region to Nova Scotia, and south to New Mexico, Texas, and Florida. Winters in South America.

Barn Swallow *Hirundo rustica*

Found almost throughout the world, the Barn Swallow is familiar and widespread in most of North America. It almost always locates its mud nests on man-made structures, usually choosing buildings that afford some protection from the elements. Extremely graceful in the air, these swallows seem always to be on the move, diving and turning swiftly, and gliding only infrequently.

Identification 5¾–7¾″. Blue-black above with rusty or buff underparts (paler in female); tail very long and deeply forked, with white spots near base below. Forehead and throat usually a deeper rust than breast.

Voice A continuous series of twittering and chattering notes; also a liquid *slip-lip*.

Habitat Open areas, farmland, marshes, lakeshores, and suburban areas; usually near water.

Range Breeds from Alaska, N. British Columbia, and N. Alberta through S. Canada to Nova Scotia, and south in most of United States; absent from Gulf Coast and parts of the Southeast. Winters in South America.

60

Red-breasted Nuthatch *Sitta canadensis*

The Red-breasted Nuthatch makes its home in coniferous forests, where it gleans insects by moving headfirst in a spiral path down the trunk of a tree; it may thereby find food that has been overlooked by woodpeckers and other birds that forage upward on a tree trunk. In winter, the Red-breasted Nuthatch feeds chiefly on conifer seeds; in years when seeds are scarce, large numbers of these birds head southward.

Identification 4½–4¾″. Blue-gray above, rusty below, with paler throat; face and eyebrow white with bold black eye stripe; crown and nape black in male, grayish in female.

Voice A nasal, tinny *yank-yank* or *nyak-nyak.*

Habitat Coniferous forests; often ranging widely in winter to other habitats if conifer seeds are scarce.

Range Breeds from SE. Alaska to Newfoundland, south to S. California, Colorado, Michigan, and New Jersey; south in mountains to North Carolina. Winters in most of breeding range and south to Mexico and Gulf Coast.

62

Western Bluebird *Sialia mexicana*

This member of the thrush family usually nests in old woodpecker holes and natural tree cavities. The related Mountain Bluebird (*S. currucoides*) occurs with this species in woodland areas of the West; the male can be distinguished from the male Western by its paler blue color and all-blue breast; the female Mountain Bluebird is grayer overall than the female Western.

Identification 6–6½″. Male deep blue on head, back, and wings, with chestnut patch on back; breast and flanks chestnut; belly gray-blue. Female grayish above with pale blue wings; gray below with pale rust wash on breast and flanks. Juvenile strongly spotted, with blue in tail and wings.

Voice Song a short, soft warble; call note a *phew* and a harsher *chuck*.

Habitat Forest edges, woodlands, orchards, and pasture areas with scattered trees.

Range Breeds from S. British Columbia and Alberta to S. California and W. Texas. Winters in most of southern half of breeding range.

64

Lazuli Bunting *Passerina amoena*

Buntings are round little birds with conical, finchlike bills; they are closely related to the cardinals and the Rose-breasted Grosbeak. The Lazuli Bunting nests in thickets and low bushes, and the male sings continually from exposed perches. In the Great Plains, it sometimes hybridizes with the eastern Indigo Bunting (*P. cyanea*).

Identification 5–6″. Male has bright blue hood, back, and rump; sides and breast orange-brown; belly white; wings dark with 2 white wing bars. Female and juvenile grayish brown above, buff below, with pale wing bars and some pale blue on wings, rump, and tail; juvenile may have streaked breast.

Voice Song a loud series of sweet, jumbled, rising and falling notes, some repeated. Calls a *pit* or *chip* and a dry buzz.

Habitat Streamside thickets, brushy areas, and woodland clearings with scattered shrubs.

Range Breeds from S. British Columbia to Manitoba, south to S. California and Oklahoma; absent from Pacific Northwest. Winters from S. Arizona southward.

66

Steller's Jay *Cyanocitta stelleri*

Like many other members of the crow family, Steller's Jay does not show much fear of people, and is quick to take advantage of feeding opportunities at campsites and bird feeders. Away from human settlements, this bird is found deep in coniferous forests, where it can be shy. Throughout most of the West, this is the only jay that has a crest—a feature that makes this species easy to recognize.

Identification 13″. Black above, with black breast and deep blue belly, tail, and wings. Black crest prominent. Inland birds have white eyebrow.

Voice A loud, harsh *chook-chook-chook* or *shack-shack-shack*.

Habitat Coniferous forests; also oak and pine woodlands, especially in fall.

Range S. Alaska to S. California and W. Texas; absent from some desert areas. Nonmigratory, although some populations move downslope in winter.

68

Belted Kingfisher *Ceryle alcyon*

With its long bill and prominent crest, the Belted Kingfisher looks somewhat top-heavy. Usually solitary, it perches conspicuously out in the open, on a branch overhanging a pond or stream, and plunges headfirst into the water to catch fish or aquatic insects. These birds fly with rapid, irregular wingbeats; they sometimes hover over the water, watching their prey, before making a dive.

Identification 12–14″. Blue-gray above, white below, with long, daggerlike bill and bushy crest; white collar about neck and broad blue-gray band across breast; female also has rufous band crossing upper abdomen.

Voice A loud, long, dry rattle.

Habitat Rivers, lakes, and seashores.

Range Breeds in much of Alaska and S. Canada, south through United States; absent from deserts of the Southwest. Winters along coasts from Great Lakes through Mississippi Valley; also farther south.

70

Cedar Waxwing *Bombycilla cedrorum*

In winter, this sleek-looking bird travels in large flocks, lisping noisily. Cedar Waxwings often alight en masse in an orchard, where they make quick work of the fruit growing there; within a few hours, they move on to a different neighborhood or grove of trees and start all over again. In the West, this species sometimes occurs with the Bohemian Waxwing (*B. garrulus*), which is similar but larger and grayer overall.

Identification 6½–8". Sleek bird with soft brown upperparts and breast, and bold mask through eyes; wings gray with hard, red, waxy tips on inner feathers; belly yellowish; tail dove-gray with yellow tip.

Voice A thin, high, lisped *ssseee* or *tseee tseee tseeee*.

Habitat Orchards, residential areas, and open woodlands, especially with fruit trees.

Range Breeds from SE. Alaska and British Columbia to Newfoundland, south to N. California, Virginia, and in mountains to Georgia. Winters through most of United States, but absent from high mountains.

Common Yellowthroat *Geothlypis trichas*

Common Yellowthroats spend much of their time near the ground in thickets or brush that offer good cover. Because they breed almost throughout North America, these birds show a fair amount of geographical variation; as a rule, birds in the West tend to be brighter yellow than their eastern counterparts. At nesting season, the male can be conspicuous, flying up from a perch to deliver a musical medley of notes.

Identification 4½–6″. Greenish-brown or olive above with bright yellow breast and throat. Male has broad black mask across eyes; female and immature lack mask.

Voice A fast, repeated *witchity-witchity-witchity-wit;* call note a *tchip* or *chik.*

Habitat Brushy swamps, moist brambles, and thickets, streamside growth, grassy marshes, and forest edges near water.

Range Breeds throughout most of North America; winters from South Carolina and central California southward.

74

Western Meadowlark *Sturnella neglecta*

This cheerful songster of open areas is nearly identical to the Eastern Meadowlark, with which it overlaps in parts of the Great Plains and the Southwest. Despite their similarity, these birds only rarely interbreed. Like the true larks (to which they are not related), meadowlarks often sing while flying up from the ground. John James Audubon named this species *neglecta* because it had been overlooked by Lewis and Clark on their expedition.

Identification
8–10". Mottled buff, brown, and black above and on flanks; bright yellow below and on sides of face, with broad black V on upper breast. Tail has white outer feathers. Bill straight and pointed.

Voice
Song a flutelike series of whistles followed by a jumble of musical notes.

Habitat
Grasslands, savannas, and meadows.

Range
Breeds from S. British Columbia to Great Lakes region, south to Mexico, W. Texas, Nebraska, and Illinois. Winters in southern part of breeding range, east to Mississippi Valley.

76

Yellow Warbler *Dendroica petechia*

This little yellow bird is a familiar sight in suburban and residential areas, where it frequents ornamental shrubs in the landscape. In wilder areas, it is often seen in willow and alder thickets. It is widespread and shows much geographical variation, and as many as seven races have been distinguished in North America; one very pale form occurs in the desert Southwest.

Identification 4½–5". Bright yellow below, yellow-green above, with 2 bold yellow patches in tail; male has thin rusty streaks on breast; usually somewhat brighter than female. Immature female olive-green; immature male resembles adult female.

Voice A cheerful, musical *sweet-sweet-sweet*, *sitta, sitta, see*; also a soft, distinctive *chip* call.

Habitat Woodlands and thickets, especially near streams; often in alder and willow thickets, gardens, and swampy areas.

Range Breeds from Alaska through most of Canada and United States; absent from tundra and from parts of the South. Winters in the tropics.

Evening Grosbeak *Coccothraustes vespertinus*

This chunky grosbeak was originally a bird of the West, but in recent years has expanded its range eastward. It nests in coniferous forests, where with the help of its huge conical bill it pries up the bracts of pine cones to get to the seeds within. At other times of year it frequents deciduous forests and is a common visitor to bird feeders, where it feasts on sunflower seeds.

Identification 7–8½". Stocky, with large, greenish-yellow, conical bill. Male has brown head with bold yellow forehead; back and belly bright yellow; wings black with large white patches. Female much grayer, with little yellow.

Voice Song a wandering series of musical whistles; call note a loud, ringing *cleep.*

Habitat Coniferous and mixed forests; a variety of habitats after breeding season.

Range Breeds from S. British Columbia and Alberta south through western mountains to N. California, Nevada, and S. Arizona; also eastward in the North to Minnesota and New England. Some birds move south in winter.

80

Western Tanager *Piranga ludoviciana*

The tanagers are a very large group of brightly colored, mainly tropical species of the Western Hemisphere; only five occur in North America. Like the other tanagers in our range, the Western Tanager looks somewhat like an oriole, but has a more conical bill. It is one of the most colorful birds of the western mountains, but it can be hard to observe; it is most easily located by voice.

Identification 7". Breeding male has bright red head, yellow throat, nape, and belly, and black wings with 1 yellow wing bar and 1 white wing bar. Female, immature, and winter male paler lemon or olive-yellow with dull olive to gray back and wings, and 2 narrow wing bars.

Voice A loud series of 2- and 3-syllable notes, separated by pauses; call a *pit-err-ick.*

Habitat Coniferous forests; less commonly in mixed deciduous forests and pinyon-juniper woodlands.

Range Breeds from S. Alaska and N. Alberta south to S. California and extreme W. Texas; absent from some desert areas. Winters mainly from Mexico south.

Northern Oriole *Icterus galbula*

In the West, this species is known as Bullock's Oriole; until recently, Bullock's was believed to be a different species from the eastern Baltimore Oriole. For many years divided by the treeless Great Plains, the two birds developed different coloration; today, they meet along wooded streams and farmland, where they interbreed.

Identification
7–8½". Male orange-yellow below with black eye stripe, crown, chin stripe, back, and central tail feathers; wings black with prominent white patch. Cheeks and eyebrow orange. Female has yellow hood, throat, and upper breast; wings pale gray with faint white patch; belly gray. Immature male like female, but has black on face.

Voice
A loud whistled series of *wheew, wheew, wheew* notes, interspersed with clucks. Also a chattering call.

Habitat
Deciduous forests, woodlands, agricultural areas, city parks, and suburbs; in winter, visits eucalyptus trees.

Range
Bullock's breeds from S. British Columbia and Saskatchewan to Mexico and Texas; Baltimore Oriole in East. Winters primarily in the tropics.

84

American Robin *Turdus migratorius*

An eloquent singer, the American Robin may be the best-known of all North American birds. In early April and May, the Robin greets the dawn with an energetic chorus of cheerful notes, as if to welcome the return of warm weather and long, sunny days. In residential areas this species can be quite tame, but populations of the forests are rather shy.

Identification 9–11". Dark gray-brown above, with brick-red or orange breast and belly; head and tail blackish, throat white. Juvenile gray-brown above, pale orange with black spots below.

Voice A rich, loud song of rising and falling phrases: *cheerily, cheerily, cheer-up, cheer-up.* Also a loud *weep* note and a lisped *see-lip* in flight.

Habitat Woodlands, forests, gardens, and suburban backyards.

Range Breeds throughout most of North America; winters mainly in southern two-thirds of United States; southernmost populations may be nonmigratory.

Black-headed Grosbeak *Pheucticus melanocephalus*

At first glance this orange-and-black species looks like a plump oriole, but it can be easily distinguished by its stout bill. Retiring and somewhat secretive, the Black-headed Grosbeak is not at all closely related to the Evening Grosbeak, although both have a large, conical bill. The Black-headed is an early fall migrant, sometimes starting its journey southward in July.

Identification 7–8½". Male orange below and on neck, with all-black hood; wings and tail black with bold white spots; lower belly and wing linings yellow. Female and immature buff below with thin streaks on flanks; dark brown wings and tail have faint white markings. Bill conical, thick, and pale in all plumages.

Voice Song a rising and falling robinlike series of fluty whistles. Call a hard *spik*; also a plaintive *whee* or *whee-you*.

Habitat Deciduous forests, streamside groves, gardens, and orchards.

Range Breeds from S. British Columbia and Saskatchewan to S. California and Mexico. Winters mainly in Mexico.

Rufous-sided Towhee *Pipilo erythrophthalmus*

The western form of the Rufous-sided Towhee was formerly thought to be a distinct species; it was known as the Spotted Towhee for the bright white spots on its upper wings and back. Towhees kick both feet backward simultaneously, raking through the leaf litter to search for insects and seeds. This noisy activity frequently calls attention to the little bird's presence in the undergrowth.

Identification 7¼–8¾". Western male has black head, breast, and back; wings black with white spots; flanks and sides rufous; central belly white. Female similar, with slightly paler head, breast, and upperparts. Immature brownish; head, nape, and back washed with rusty tones.

Voice Song a *chup-chup-zeee* or *chew-wee*; call a slurred *tow-hee?*

Habitat Woodlands, forest edges, gardens, and parks with low shrubby growth; avoids dense forests and treeless plains.

Range Breeds from S. British Columbia to Maine, south to Mexico and Florida; absent from much of Great Plains. Winters in southern two-thirds of United States.

Dark-eyed Junco *Junco hyemalis*

Four forms of the Dark-eyed Junco occur principally in the West. Once considered separate species, they are all conspicuous ground-feeding sparrows with pink bills and white outer tail feathers. By far the most common and widespread is the Oregon Junco, shown here. The White-winged is gray with two white wing bars; it is found mainly in the Black Hills. The Pink-sided has broad pink sides and flanks, and occurs in the eastern Rockies. The Gray-headed has a reddish-brown back and light gray head; it is most common in the Southwest.

Identification 5–6½″. Male Oregon Junco has black hood, brown back, and dark wings with white edges; belly whitish; flanks buff. Female similar but with paler hood. Pink bill and white outer tail feathers in all forms.

Voice A trill with only occasional changes in pitch.

Habitat Forests, woodlands, brushy areas, and lawns.

Range Oregon Junco breeds from British Columbia to NW. Montana and south. Winters mainly along coast. Other forms have more restricted ranges. Also in the East.

Mountain Chickadee *Parus gambeli*

These energetic little birds are abundant in the conifer forests of the western mountains, where they nest in tree cavities. They travel in flocks, sometimes in company with warblers and vireos, and may range far downslope in cold weather. At such times they often visit feeders, where their boldness is rewarded with sunflower seeds.

Identification 6". Gray with black cap, eye stripe, and throat; cheeks and eyebrow white to pale gray. Birds of Great Basin area paler with more buff; birds from farther west darker.

Voice A hoarse *chick-a-dee, dee, dee* and a whistled *fee-bee* or *dee-dee.*

Habitat Montane conifer forests; occasionally elsewhere in winter.

Range British Columbia south through western mountains to Baja California and SW. Texas. Usually nonmigratory, but some birds move downslope in winter.

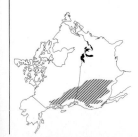

94

Yellow-rumped Warbler *Dendroica coronata*

Widely known as Audubon's Warbler, this bird is actually the same species as a white-throated eastern bird, the Myrtle Warbler; technically, both forms are now known as the Yellow-rumped Warbler, although many birders prefer to use the old names. The summer ranges of the two forms overlap in southwestern Canada, where the birds hybridize.

Identification 5–6". Breeding male slate-blue above with darker streaks; rump bright yellow; charcoal-gray patch around eyes; throat yellow; small, bright yellow patch on crown and sides. Female, immature, and nonbreeding male mainly gray-brown, similar to breeding male but duller; with yellow rump and white spots in tail.

Voice Song a thin, musical trill, *twee-twee-twee*; call a soft *chep*.

Habitat Coniferous and mixed forests, often in mountains.

Range Audubon's breeds from central British Columbia and Saskatchewan south in mountains to Mexico. Winters from S. British Columbia along coast to Mexico; also in the Southwest to Texas. Myrtle has eastern range.

Golden-crowned Kinglet *Regulus satrapa*

The species name *satrapa* comes from Greek, and it means the ruler of a small province or a petty official. This spry little bird often flicks its wings nervously as it combs its little kingdom of conifer trees, searching for insects and larvae. The Golden-crowned Kinglet is frequently seen with its relative, the Ruby-crowned Kinglet (*R. calendula*), which is similar but has a red tuft of feathers on its crown.

Identification 3½–4". Olive above with 2 white wing bars; paler grayish white below; white eyebrow; bill slender; tail has small notch. Crown deep gold or orange-yellow in male, yellow in female, and bordered with black in both sexes.

Voice A thin, rising *see-see-see*; also a *tseeep*.

Habitat Conifer forests in nesting season; other woodland habitats as well in winter.

Range Breeds from Alaska to Newfoundland, south in suitable conifer-forest zones to California, Colorado, Minnesota, and southern Appalachians. Winters from southernmost Canada throughout most of United States.

Pine Siskin *Carduelis pinus*

Both the common and scientific names reflect this species' reliance on pine seeds, although these birds cheerfully consume the seeds of hemlocks, cedars, and other evergreens. Pine Siskins occur in noisy flocks and are frequent visitors to feeding stations, especially when conifer seeds are scarce. Near feeders and birdbaths, these little birds can be quite tame and approachable.

Identification 4½–5¼". Brown above and below with darker streaks; some birds are paler overall, others darker; yellow on wing and on deeply notched tail. Females have less yellow.

Voice A harsh *shick-shick* and a buzzy, ascending *bzzrrreeee.* Call note a *sweeeeeet.*

Habitat Conifer forests; also in alders, aspens, and other deciduous trees near northern conifer forests.

Range Breeds from S. Alaska to Newfoundland south through most of the forested West; east through Great Lakes region to New England. Winters irregularly farther south.

100

White-crowned Sparrow *Zonotrichia leucophrys*

There are several different races of the White-crowned Sparrow, breeding from northern Alaska to California. When the time comes to migrate, the northernmost races travel farthest; races with progressively more southern breeding ranges move shorter and shorter distances. Those that live near San Francisco hardly move at all.

Identification 5½–7". Adult has bold black-and-white head pattern: crown white with broad black stripe; eyebrow white; eyeline black. Bill pink to yellow. Face, neck, and breast gray; back and wings streaked black and brown, with white wing bars. Immature has brown and buff head stripes.

Voice Song a series of whistles, repeated or followed by a trill; varies regionally. Call a *chink* or *seet*.

Habitat Forest edges, bogs, meadows, parks, and suburbs.

Range Breeds from N. and W. Alaska east to N. Quebec and south in the West to central California and N. New Mexico. Winters along Pacific from N. British Columbia to Baja California and through most of the South.

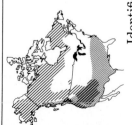

102

Chipping Sparrow *Spizella passerina*

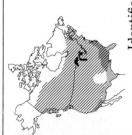

At close range it is easy to see the bright rufous cap of the compact little Chipping Sparrow. The bird gets its name from its song, a series of "chip" notes, sometimes run together in a fast trill. Common in residential areas, this sparrow is very tame; it often visits feeders, and may take crumbs from a person's hand.

Identification 5–5¾". Adult has black forehead, rusty crown, white eyebrow, and black eye stripe. Upperparts streaked in brown and black, with 2 white wing bars; underparts, cheek, and back of neck clear gray. Immature buffier, more streaked, without bold black and white markings or rusty crown.

Voice A thin, insectlike trill on 1 note. Call a sweet, high *tseep*.

Habitat Forest edges, orchards, brushy pastures, city parks, and gardens.

Range Breeds throughout most of Canada and south throughout United States to S. Arizona, New Mexico, and S. Texas; absent from most of Florida. Winters from S. California, S. Texas, and Maryland southward.

Song Sparrow *Melospiza melodia*

This common and abundant sparrow shows a wide range of geographical variation. There are 34 recognized races, most with the same basic pattern of brown-and-white plumage. One race from the Aleutian Islands, however, looks almost like a different species altogether: It is much larger and darker than the typical form, with a longer bill.

Identification 5¾–7". Brown above with grayish streaks; white below with heavy brown streaking and large spot at center of breast. Tail usually has more reddish brown than back.

Voice Song has 3 sweet notes followed by a lower note and a trill. Call note a *chimp.*

Habitat Thickets, forest edges, marshes, gardens, and city parks.

Range Breeding range from S. Alaska through S. Canada to Newfoundland; south to S. California, New Mexico, NE. Kansas, and North Carolina. Winters from southern half of breeding range to Mexico and Florida.

Lincoln's Sparrow *Melospiza lincolnii*

Lincoln's is closely related to the Song Sparrow, but is more furtive than its cousin. This bird skulks through brushy undergrowth, searching for insects and seeds; like the towhees, it rakes through the leaf litter by kicking backward with both feet at once. Birders sometimes tempt Lincoln's Sparrow out of its brushy cover by loudly kissing the back of the hand—an imitation of a bird in distress.

Identification 5¼–6". Gray with brownish streaks above; eyebrow and sides of neck gray. Buff below, with fine dark streaks. Immature resembles immature Song Sparrow, but more finely streaked below.

Voice Song a gurgled melody, rising and then falling. Call a *tik* or *tsup*.

Habitat Wet meadows and bogs with brush; in migration, weedy fields, willow thickets, and gardens.

Range Breeds from N. Alaska to Newfoundland, south in mountains to S. California and N. New Mexico. Winters from central California to N. Alabama and south.

American Dipper *Cinclus mexicanus*

The only aquatic songbird in North America, the Dipper lives in the mountains of the West, close to cold, rushing streams. Its wrenlike song is given all year, especially when streams are very full; it can be heard even over the noise of the water. From the bank or a boulder, the Dipper plunges into the icy water to take water striders, mosquito larvae, and other insects. Dippers often bob up and down on a rock; when alarmed, they fly away, keeping low over the water.

Identification 7–8½". Stout, wrenlike; slate gray with long, pointed bill and stubby tail. Immature and winter adult may be paler, especially on outer wings and below.

Voice Song a loud, rich, musical medley of trills and runs. Alarm note a sharp *bzeet*, given in flight.

Habitat Near fast, clear, rushing mountain streams.

Range Breeds from Alaska to S. California and New Mexico; does not migrate, but may move downslope in very cold weather.

110

House Sparrow *Passer domesticus*

This common little bird is easy to recognize and is often one of the first songbirds learned by beginning birders. Introduced to North America in the middle of the 19th century, the House Sparrow is now found almost throughout the continent; it thrives in association with people, and is particularly abundant in cities and farm areas. Like many introduced species, it competes vigorously with native birds for nesting sites.

Identification   5½–6¼". Male streaked above with brown and black, with white wing bar; throat and upper breast black; nape chestnut and crown gray, with chestnut line through eye. Female streaked brown and black above, dingy gray below, with dull stripe behind eye.

Voice   A repeated *chirp, cheep,* and various twitters.

Habitat   Farmland, cities, towns, and suburban areas.

Range   Throughout most of S. Canada and entire United States in cities, towns, suburbs, and agricultural areas.

House Finch *Carpodacus mexicanus*

Abundant throughout the West, the House Finch nests in tree cavities or the old nests of other birds; in cities and residential areas it seems content to dwell in tin cans, building ledges, and air conditioners. House Finches also occur in the East, but the ones there are descended from caged birds that were trapped and sold illegally (as "Hollywood Finches") until wildlife authorities stopped the trade.

Identification 5–5½". Male pale brown with darker streaks, and with bright red on forehead, eyebrow, breast, and rump. Female similar but lacks red, and has uniform brown head.

Voice Song a clear, canarylike warble, ending in an ascending *zeeee*. Call note a chirp.

Habitat In the East, in cities and residential areas; in the West, in desert scrub and chaparral.

Range S. Canada to S. Mexico, east to Idaho, Nebraska, and central Texas. Introduced in the East. Does not migrate.

114

Rufous Hummingbird *Selasphorus rufus*

This aggressive little nectar-feeder times its spring arrival in Oregon to coincide with the blooming of a red-flowered currant that is one of its chief food sources. The male and the female have separate territories, and both are combative, chasing away other hummingbirds, blackbirds, and thrushes. The male performs a circular courtship flight, rising high in the air over the branch where the female is perched.

Identification 3¼–3½". Adult male orange-rufous above, with bright orange-red throat patch (gorget) and white breast. Female green above with rufous sides and base of tail. Immature male has green back, red flecks on throat.

Voice Gives low, soft *chuppy* or *chippy* notes; also an excited *zeee-chuppity-chup*. Male's wings produce buzzy trill in courtship flight.

Habitat Woodlands, forest edges, chaparral, and mountain meadows.

Range Breeds from SE. Alaska to S. Oregon, east to SW. Alberta and Montana. Winters in Mexico.

116

House Wren *Troglodytes aedon*

Despite its small size, the House Wren is aggressive in competition for nesting sites. It sometimes tosses the nest, eggs, and even the young of other cavity-nesters out onto the ground. In early spring, the male returns early from the south and builds several rough "dummy" nests; when the female arrives, she may use one of these establishments or start a new nest altogether.

Identification 4½–5¼". Small and plump. Dull brown above, with faint dusky bars on wings; grayish white below. Tail short, with dark bars and no spots.

Voice Song a rising and falling, bubbling chatter, repeated many times.

Habitat Woodland edges, farms, city parks, and residential areas.

Range Breeds from S. British Columbia, N. Alberta, Ontario, and Maine south to S. California, Arizona, N. Texas, and Georgia. Winters from S. California to Gulf Coast and South Carolina southward.

Cactus Wren *Campylorhynchus brunneicapillus*

The largest wren in North America, this bird is a denizen of the desert Southwest. It may often be seen atop a spiny cactus, with head up and tail pointing down, delivering its harsh, low song. It nests in chollas and yuccas, whose prickles and sharp leaves keep would-be intruders at bay.

Identification 7–8¾″. Pale brown above, with bold black-and-white patterning on back, wings, and tail; buff below with heavily spotted throat and upper breast; flanks and belly have less heavy dark spots. Crown rust-brown; eyebrow white. Tail long; bill long, slightly downcurved.

Voice Song a low, gravelly *chug-chug-chug-chug* or *cora-cora-cora.*

Habitat Low-elevation deserts with cactus, palo verde, mesquite, and other thorny vegetation.

Range S. California to S. Texas, south into Mexico; does not migrate.

120

Northern Flicker *Colaptes auratus*

In the West, there are two forms of the Northern Flicker, both once considered separate species. The Gilded Flicker, a bird of the deserts, has a brown crown and yellow wing linings; otherwise, it looks and behaves like this bird, the Red-shafted Flicker. A third form, the Yellow-shafted Flicker, is common in the East and sometimes visits the West in winter. Flickers are the only woodpeckers that frequently feed on the ground.

Identification   11–14″. Brown above with dark spots and bars; buff-white below with black spots and with black patch on upper breast; face gray, with red mustache; pinkish-orange wing linings and white rump patch visible in flight.

Voice   A loud, repeated *wik-wik-wik* or *flicker-flicker-flicker;* also a loud *kleer.*

Habitat   Woodlands and forests; Gilded Flicker in deserts.

Range   Breeds throughout North America to northern limit of trees; Red-shafted and Gilded principally in the West; some northern populations move south in winter.

122

Downy Woodpecker *Picoides pubescens*

The Downy is the most common and familiar woodpecker in most of North America. In the West, it occurs primarily in groves along the banks of streams, although it is also a frequent visitor to city parks and feeders in residential areas. In fall, the Downy often travels with flocks of other little birds, mainly chickadees, kinglets, and nuthatches.

Identification 6–6½". Black and white above, white below; white cheeks intersected by black eyeline; thin mustache runs from bill to back of neck. Male has small red patch on nape.

Voice Call note a dull *pik*. Also gives a loud, descending rattle. Drums with its bill against bark, producing fast series of percussive noises.

Habitat Forests, woodlands, orchards, residential areas, and city parks.

Range Alaska through most of southern half of Canada, and throughout most of United States; absent from treeless deserts. Some northern birds move south in winter.

124

Acorn Woodpecker *Melanerpes formicivorus*

The sociable Acorn Woodpecker is found in oak and oak-pine woodlands in the West, where it feeds on acorns; in towns and residential neighborhoods, it also consumes walnuts, pecans, and other delicacies. In the fall, the Acorn Woodpecker carefully stores acorns in small holes it drills in the trunks of trees, wedging the nuts in tight to keep them safe from squirrels.

Identification 8½". Adult male black above with red crown patch; black extends over eye; forehead, cheek, and throat creamy white; upper breast black; belly, wing patches, and rump white. Female similar but with black forecrown.

Voice A loud *Ja-cob, Ja-cob* or *ya-cup, ya-cup, ya-cup*; also drums with bill.

Habitat Pine-oak and oak woodlands, city parks, and suburbs.

Range S. Oregon through W. California to Baja California; Arizona, New Mexico, and W. Texas. Does not migrate.

126

Red-winged Blackbird *Agelaius phoeniceus*

The familiar Red-winged Blackbird forms large flocks with the similar Tricolored Blackbird (*A. tricolor*) of western California and southern Oregon. The red shoulders of the Tricolored are bordered with white, and the bird has glossier plumage and a thicker bill than its cousin. Both birds nest in marshes; they frequent farm areas and open country after the breeding season.

Identification   7½–9½". Male black with bright red shoulder patches. Female and juveniles have heavy dusky brown streaks.

Voice   Song a liquid, musical *ob-ka-lee!* Also various *chuck* and *kink* notes.

Habitat   Usually nests in marshes and other wetlands, especially areas with cattails; also in moist thickets, pastures, and meadows.

Range   Breeds from S. Alaska, N. Alberta, and Ontario to Maritime Provinces, south through entire United States. Winters in southern two-thirds of United States, including the temperate Northwest; absent from southern Appalachians.

European Starling *Sturnus vulgaris*

Introduced from Europe just before the turn of the century, the adaptable Starling is now ubiquitous; it lives in a wide variety of habitats, from crowded cities to agricultural areas, and has even become somewhat common in parts of the Southwest, where it has only recently arrived. It flies in dense flocks that wheel and turn in unison, and large numbers often gather to form huge roosts.

Identification 7½–8½". Short-tailed, chunky. In spring, black with iridescent greenish gloss; bill yellow. Winter plumage heavily flecked with white; bill dark. Immature dusky gray-brown above, paler below.

Voice A wide variety of squeaks, chattering notes, whistles, and clicks; often gives a "wolf-whistle" and mimics other birds as well.

Habitat Cities, parks, orchards, woodlands, and farm areas.

Range Throughout United States and S. Canada.

Brown-headed Cowbird *Molothrus ater*

Cowbirds often congregate with other members of the blackbird family near farms and rangeland, frequently in close association with livestock. Like the related Bronzed Cowbird (*M. aeneus*), which occurs only in southern Arizona and south Texas, the Brown-headed is a brood parasite; it makes no nest of its own, but lays its eggs in the nests of other species. The young cowbird is usually so much larger than the nestlings of its songbird host that those little birds may starve or be crowded out of the nest.

Identification   6–8". Male iridescent greenish black with deep brown head; females gray-brown; juvenile gray-brown with faint streaks on breast and scaly-looking upperparts. Bill black and finchlike in all plumages.

Voice   Voice and calls squeaky, bubbly, and high-pitched; female chatters.

Habitat   Woodlands, farmlands, fields, and suburbs.

Range   Breeds throughout most of United States and S. Canada, extending into N. Alberta. Winters in S. United States.

132

Brewer's Blackbird *Euphagus cyanocephalus*

This sociable bird is commonly seen with Brown-headed Cowbirds and Red-winged Blackbirds. Brewer's eats a variety of insect pests, such as aphids, cankerworms, grasshoppers, termites, and weevils, and often follows along behind a tractor as it plows the soil. It nests from low-elevation farmland to mountain meadows.

Identification 8–10". Breeding male glossy black with a long, sharp, conical bill and a whitish eye. In good light, head looks purplish and body looks greenish. Female and immature gray-brown, darker on lower back, wings, and tail; eyes brown.

Voice Song a hoarse, creaky *kseeee* or *ksheek;* call a harsh *check!*

Habitat Pastures, riverside thickets, brushy savanna, towns, farms, and ranches.

Range Breeds from central British Columbia east to Michigan, south to S. California, Nevada, and New Mexico. Winters from S. British Columbia, Colorado, Oklahoma and S. Carolina south into Mexico.

134

Common Raven *Corvus corax*

Large, imposing, and entirely black, the Common Raven is often seen soaring and gliding in mountainous country. It frequently flies over roads and highways, looking for carrion, although it is omnivorous and an opportunistic feeder. In flight ravens can be distinguished from the related American Crow by their longer wings and long, wedge-shaped or rounded tail; a crow's tail in flight looks squared.

Identification 21½–27". Large and stocky; all-black with heavy black bill; neck feathers shaggy. Rounded wings and wedge-shaped tail visible in flight.

Voice A low, hoarse, croaking *crock* or *kraaak*.

Habitat Mountains, deserts, beaches, forests, and Arctic tundra.

Range Alaska and Canadian Arctic to Newfoundland, south through British Columbia and W. Alberta to S. California and W. Texas; in East, south to Great Lakes region and N. New England, and in Appalachians.

American Crow *Corvus brachyrhynchos*

Common throughout much of North America, this large, black bird is a familiar sight in cities and towns as well as in open country and along the seashore. Crows are intelligent, and captive birds tested in puzzle-solving have provided many valuable insights into the learning process. American Crows are gregarious birds, and they form large flocks; they are sometimes joined by a handful of ravens.

Identification 17–21". Large and stocky. Black all over, with slight purplish sheen. Bill stout; tail squared.

Voice A raucous, familiar *caw, caw, caw.*

Habitat Open areas, woodlands, fields, suburbs, orchards, gardens, and city parks; tends to avoid deserts and dense forests.

Range Breeds throughout southern two-thirds of Canada and most of United States; winters south of Canadian border. Absent from much of interior Southwest.

138

Black-billed Magpie *Pica pica*

The large, black-and-white Black-billed Magpie has a remarkable tail—longer than its body—that streams behind the bird in flight. This species is similar to the Yellow-billed Magpie (*P. nuttalli*), which occurs only in central and southern California. The latter can be distinguished by its yellow bill and a patch of bare yellow skin just below the eye. Black-billed Magpies, like most members of the crow family, are omnivorous scavengers.

Identification
: 18–22". Black above and white below, with white wing patch and shoulders visible in flight. Tail long; appears tapered in flight. Bill dark.

Voice
: A high, nasal *mag?* and a harsh *check-check-check*.

Habitat
: Open countryside, savanna, brushy areas, and streamside thickets.

Range
: Breeds from S. Alaska through British Columbia and east to Manitoba, south to east-central California and Nebraska. Mostly nonmigratory; some wander north and east of range in fall and winter.

Rock Dove *Columba livia*

The Rock Dove—alias the Pigeon—is thoroughly adapted to living in association with people. Native to Europe, it was probably the first bird species to be domesticated; a homing pigeon was reportedly used to carry the news of Julius Caesar's conquest of Gaul back to Rome. There are some Rock Doves that live in the wild, nesting as their ancestors did on rocky sea cliffs; these wild birds are notably shy of humans.

Identification 12–13". Stocky, with short, fan-shaped tail. Birds in the wild (and many in cities) bluish-gray with 2 narrow black wing bars, white rump, and some iridescence on sides of neck. Color variations include all-black, all-white, piebald, and reddish-brown birds.

Voice A soft, rolling *coo-croo* or *coo-took-crooo.*

Habitat Cities, parks, gardens, suburbs, and farm areas; rocky canyons or sea cliffs.

Range Throughout S. Canada and United States; does not migrate.

142

Band-tailed Pigeon *Columba fasciata*

Larger than the related Rock Dove, the Band-tailed Pigeon is a shy forest bird, although in some places it is beginning to venture near feeders and backyard berry sources. In the early part of this century the species was hunted almost to extinction, but it seems to have recovered. The two North American subspecies are distinguished chiefly by habitat: One is found in humid forests along the coast, and the other occurs in drier mountain woodlands of the interior.

Identification
13–15″. Dark gray above; long tail has wide, pale gray band at end. Head and underparts purplish or pink in male, gray-brown in female. Both sexes have whitish crescent on nape.

Voice
A low, owl-like *whoo-hoooo*; often repeated.

Habitat
Moist conifer forests along coast; pine-oak woodlands inland.

Range
SE. Alaska along coast to Baja California; Utah and Colorado south to Central America. Winters from Washington, S. Arizona, and S. New Mexico south.

144

Mourning Dove *Zenaida macroura*

The Mourning Dove is hunted extensively in some parts of North America, and it is estimated that more than half of the young birds born each spring do not live beyond their first year. Fortunately, these birds raise two to four broods each season, and so the Mourning Dove is still common. The sad notes of the bird's song are usually heard just before the dawn in spring and summer.

Identification 11–13". Slim, with small head and long, tapered tail. Soft, sandy brown or brownish gray above, with a few black spots; paler below, sometimes washed with pale cinnamon; tail feathers tipped with white.

Voice A low, sad *whoo-oo, hoo, hoo, hoo*; second note rises sharply.

Habitat Almost anywhere except dense forests: woodlands, streamsides, desert washes, gardens, city parks, and suburban backyards.

Range Breeds from S. Alaska and W. British Columbia through S. Alberta and Great Lakes to New England; south to Mexico and Florida. Northern populations migratory.

California Quail *Callipepla californica*

The California Quail is a plump little ground bird with a distinctive feathery plume adorning its forehead. Common in a variety of habitats, it is very gregarious, and in winter may form large, well-organized coveys composed of several families. As the members of the covey feed, one bird acts as a sentinel, keeping a lookout for danger.

Identification   9½–10½". Blue-gray above with brownish wings; belly white with scaly pattern. Male has brown head with buff on forehead, bold white eyebrow, black face and throat, and white crescent on cheek; large, dark, fluffy head plume. Female has reduced pattern on head and face, and smaller plume.

Voice   A loud *ca-ca-cow* or *ca-caah-co*; also various clucking notes and a *pit* or *whit-whit*.

Habitat   Open areas with brush; chaparral, riverside areas, coastal canyons, and suburbs.

Range   Resident from S. Oregon to Baja California; introduced and established from S. British Columbia south.

Greater Roadrunner *Geococcyx californianus*

This large, ungainly-looking cuckoo runs swiftly through deserts and woodlands in pursuit of lizards and other prey; it also runs to escape enemies (including the coyote), and it has been clocked at 15 mph, although its usual pace is somewhat more sedate. At dawn, the male Greater Roadrunner can be heard calling through the desert stillness from the top of a mesa or the branches of a dead tree.

Identification
20–24". Large, upright, with bushy crest and long tail. Streaked brown and buff above and on breast; dingy white on belly. Iridescence in tail and upperparts visible at close range; also blue-and-red patch behind eye. Immature lacks eye patch.

Voice
A long, sad, descending *coo, coo, coo, ooh, ooh, ooh*; also a variety of whines and clucks.

Habitat
Open, dry areas with scattered brush and thickets; open woodlands, agricultural areas, and grasslands.

Range
N. California east to S. Kansas and NW. Louisiana, south into Mexico.

Ring-necked Pheasant *Phasianus colchicus*

Introduced from Asia only a little more than a century ago, the Ring-necked Pheasant is now so well established in North America that it is the state bird of South Dakota. This species runs as often as it flies, and is comfortable in a wide variety of habitats, as long as there is suitable brushy cover. A male may mate with as many as four hens, each of which establishes a nest within the male's territory or crowing area.

Identification Male 30–36″; female 20–26″. Chickenlike, with long, pointed tail. Male has bright green or blue-green head, red face, white neck-ring; body and wings iridescent bronze, gold, and red, with bold, dark spotting. Female soft brown, spotted and barred with black.

Voice A loud, crowing *cuck-cuck* or *caw-caw*, accompanied by loud wingbeats. Male cackles when taking off.

Habitat Grassy and brushy areas near woodlands; farms, pastures; also in cattail marshes in winter.

Range Much of United States and S. Canada in suitable farm regions; absent from high mountains and deserts.

Turkey Vulture *Cathartes aura*

Large and mainly black, with a naked red head, the Turkey Vulture has more appeal in flight than it does when perched ominously in a tree. When it flies, this bird holds its wings in a shallow V; it does not flap its wings often, but tilts and glides, taking advantage of rising currents of warm air to gain altitude.

Identification
26–32". Large and blackish, with small, unfeathered red head and stout bill with sharply hooked tip. Legs and feet orange. Silvery-gray wing linings conspicuous in flight, making wings appear two-toned. Immature has dark head and gray feet.

Voice
Generally silent; utters hisses and groans at nest or when disturbed.

Habitat
Dry, open country; often along roadsides; sometimes roosts in woods.

Range
Breeds from S. British Columbia to S. New England, south to Mexico. Winters from New Jersey to Florida and E. Texas; also in parts of the Southwest.

Bald Eagle *Haliaeetus leucocephalus*

Despite its fame, the Bald Eagle is rare. The victim of poisoning by pesticides, chiefly DDT, the species suffered a catastrophic decline; the outlawing of some poisons has helped to maintain these birds. The Bald Eagle is also shot illegally in many places, and in some areas its wetland habitat has been destroyed. Today, large numbers of these magnificent birds are found only in carefully patrolled wildlife refuges.

Identification   30–43". Very large, brown, hawklike bird with white head and tail and stout, hooked yellow bill. Immature variable, but with dark head and tail and black bill.

Voice   A series of squeaky, thin cackling or chittering notes.

Habitat   Seacoasts, lakes, rivers, and marshes.

Range   Breeds in forested areas of Alaska and Canada south to Oregon, N. Idaho, Great Lakes area, and N. New England; also locally along Atlantic and Gulf coasts and in Florida. Winters from S. Canada south, especially along major river systems of the interior.

156

Golden Eagle *Aquila chrysaetos*

The majestic Golden Eagle often soars with its wings horizontal, watching from high in the air for the movement of a small mammal, snake, or turtle. Now protected by law, this species was formerly killed in great numbers, partly because of a mistaken belief that the bird is a threat to livestock. Golden Eagles rarely attack a healthy large animal, but they will take crippled or ailing lambs, deer, and waterfowl.

Identification 30–40". Adult brown or dark brown overall; crown and nape edged with gold or tan. Immature similar; white wing patch and tail band visible in flight.

Voice Usually silent; occasional mews and squeals.

Habitat Mountain canyons, ranchland, open countryside, forests, and tundra.

Range Breeds from NW. Alaska to N. Quebec, south to S. California, W. Texas, N. Manitoba, and Labrador. Winters from N. British Columbia and central Quebec south to southern limit of breeding range.

Red-tailed Hawk *Buteo jamaicensis*

The most widely distributed large hawk in North America, the Red-tailed is able to tolerate a broad range of habitats. There are several races of this species; the one most commonly encountered in the West is a bird of open country; it tends to have a darker head and more heavily marked underparts than eastern Red-tails.

Identification   19–25". Large; typical bird dark brown above, usually light below with dark band on belly. Tail rufous with dark band and paler tip. Geographical variations include a dark brown color phase and a pale phase with a white tail. Immature has grayish tail with narrow bands.

Voice   A loud, harsh, descending *tseer*, usually given when disturbed.

Habitat   Grasslands, pastures, open woods, and farmlands; also in plains, tundra, and deserts with scattered trees.

Range   Breeds throughout most of North America; winters from Canadian border south.

160

Swainson's Hawk *Buteo swainsoni*

This hawk is a common bird of open areas in the West. In migration, it is often seen in huge flocks; the birds rise on warm currents of air, spiraling upward, then descend in a long glide to the bottom of another thermal air current. It frequently hunts from fence posts and low trees, and often watches for prey from the ground.

Identification 19–22″. Dark brown above with white throat and dark band across breast; outer flight feathers dark gray; wing linings white. Tail gray above, light below, with white tip. Dark-phase bird sooty-brown all over. Immature blackish brown above, buff below, with variable spots and streaks.

Voice Usually silent; sometimes whistles at nest.

Habitat Open plains, prairies, and meadows; sometimes nests in wetland areas.

Range Breeds mainly from S. British Columbia and Prairie Provinces south to south-central California, Arizona, New Mexico, and W. Texas; also in E. Alaska. Winters mainly in South America.

Northern Harrier *Circus cyaneus*

This is a common raptor of marshlands, prairies, and open, grassy areas. The Northern Harrier hunts on the wing, flying great distances every day in its search for mice and other small animals. Like owls, this bird uses its sharp hearing to locate its prey. Formerly known as the Marsh Hawk, the Northern Harrier is a skillful flyer, and the male performs a magnificent courtship display.

Identification 16–24". Slim, with long wings and tail. Male light gray above, whitish with small reddish flecks below; tail obscurely barred. Female brown above with brownish streaks below. Both sexes have prominent white rump. Immature brown above, rusty below.

Voice Usually silent; a chattering *kee-kee-kee* near nest.

Habitat Grasslands, marshes, and open fields.

Range Breeds from Alaska to N. Alberta and eastward to Newfoundland; south to S. California, N. New Mexico, Ohio, and Virginia. Winters from Washington, N. Utah, Great Lakes region, and New England south to Mexico and Florida.

164

Osprey *Pandion haliaetus*

Found throughout the world, the Osprey is usually seen near water, although a lack of suitable nesting sites often prompts it to move far inland. This bird lives almost exclusively on fish, which it takes from the water with its talons, descending on its prey like a terrestrial hawk. Its feet have tiny spicules on the bottom that help the bird to maintain its hold on slippery fish.

Identification 22–25". Large, hawklike; brown above, white below, with white head; dark brown line runs through eye and on side of face. Juvenile similar but more mottled. In flight, wings show distinctive bend at "wrist."

Voice Loud whistling and chirping given at nesting and during courtship; also a *kip kip ki-yeuk, ki-yeuk* when alarmed.

Habitat Coastal areas, lakes, and rivers.

Range Breeds from Alaska and north-central Canada to Newfoundland, south to California and Arizona, Great Lakes area, and Nova Scotia; south along Atlantic Coast to Florida and Gulf Coast. Winters along southern coasts.

Common Nighthawk *Chordeiles minor*

This species is a common insect-feeder. It typically hunts at dusk, flying with its huge mouth wide open to catch insects on the wing, but it is also seen by day, and it has adapted to a wide variety of habitats, including rooftops in cities. The Common Nighthawk's cryptically colored plumage breaks up the outline of the bird when it is perched on the ground, helping it to escape the notice of predators.

Identification 8½–10". Mottled gray, white, black, and brown above; underparts buff with brown bars. Long, pointed wings marked with white patch near bend, visible in flight, as is white throat patch. Tail long. Female slightly duller than male.

Voice A nasal, insectlike *beeerp* or *brrrrrrrp.*

Habitat Open woodlands, forests, meadows, sagebrush plains, and cities.

Range Breeds from SE. Alaska east to Quebec, south to N. California, Nevada, SE. New Mexico, Texas, and Florida. Winters in the tropics.

168

American Kestrel *Falco sparverius*

The American Kestrel often hunts on the wing, hovering over fields and open land searching for mice, lizards, small snakes, and frogs; when grasshoppers are abundant, these insects usually become the bird's chief food. Formerly known as the Sparrow Hawk, this species flies on long, pointed wings; when it lands on a perch, it often pumps its tail up and down. The American Kestrel is the smallest North American falcon.

Identification 7½–8″. Small, with long, pointed wings and rusty tail and back. Adult male has blue-gray wings and rusty crown; female has black-barred, rufous wings. Underparts white or buff, male's with black spots, female's with heavy streaks.

Voice A loud, shrill *killy-killy-killy*.

Habitat Open countryside, grasslands, farms, suburbs, and city parks.

Range S. Alaska to Newfoundland, south through South America. Winters north as far as S. British Columbia, Illinois, and New England.

Great Horned Owl *Bubo virginianus*

The magisterial Great Horned Owl gets its name from its large, conspicuous ear tufts, which it raises in moments of great excitement. These tufts are not ears at all; the powerful ears of the Great Horned Owl are hidden beneath feathers on the side of the head. Like most other owls, this species flies absolutely silently; its stealth is made possible by the loose, ragged outer edges of the flight feathers, through which the air flows without the telltale rushing sound produced by most birds in flight.

Identification   18–25″. Large with widely spaced ear tufts. Dark gray-brown with fine whitish mottling above; buff-white below, with dark brown barring and white throat. Eyes bright yellow.

Voice   A deep, sonorous, resonant series of hoots: *hoo, hoo-hoo-hoo, hoo, hoo;* or *hooo, hoo-hoo, hoooo, hoooo.*

Habitat   Forests, open country, swamps, deserts, and even large city parks.

Range   Throughout North America; usually does not migrate.

Western Screech-Owl *Otus kennicottii*

This pint-size owl was formerly thought to be identical to the Eastern Screech-Owl (*O. asio*). The two birds have distinct ranges and very different vocalizations; the western species gives a series of whistled notes that speeds up at the end. In the humid forests of the Pacific Northwest, there is a rare reddish color phase of the Western Screech-Owl.

Identification 7–11". Small, with mottled upperparts and prominent ear tufts. Underparts whitish, with streaks and bars. Some geographical variation: Birds from drier areas tend to be paler gray, those from humid areas browner.

Voice A series of 7–20 soft, whistled notes, starting slowly and speeding up, and all on 1 pitch. Also a short trill and various yelping and barking noises.

Habitat Woodlands and forests, especially streamside areas with oaks.

Range SE. Alaska south along coast; east along Canadian border to S. Alberta and N. Montana, and south to Mexico. Does not migrate.

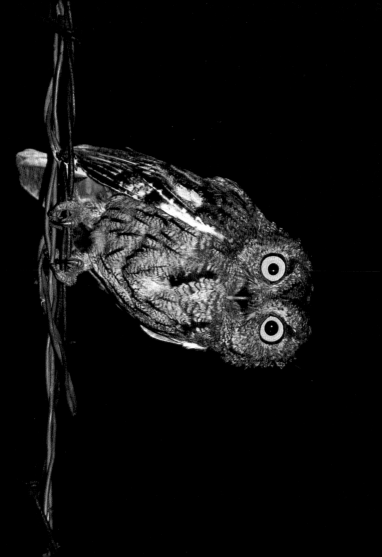

Common Barn-Owl *Tyto alba*

Although it is unmistakably owl-like, the Barn-Owl is not closely related to other North American owls, and is actually in a family of its own that includes ten other species from around the world. In flight, this large bird looks very white, especially from below. It nests in barns and other outbuildings as well as in tree cavities and caves.

Identification 14″. Golden brown above, with grayish mottling on wings and back; pale buff to white below, with sparse dark spots on breast and wing linings. Face heart-shaped, white, with long, narrow bill. Feet and legs covered with bristly white feathers.

Voice Song a long, rasping screech, increasing in volume; also gives a loud hiss.

Habitat Farm areas, marshes, prairies, and open woodlands; also suburbs and cities.

Range SW. British Columbia, South Dakota, N. Illinois, and S. New England south to Central and South America. Some northern populations move south in winter.

176

Guide to Families

Birds are arranged in groups called families and subfamilies. Knowing group characteristics is often helpful in identifying species.

Water Birds

Loons (family Gaviidae) and grebes (family Podicipedidae) are highly aquatic, diving with ease and swimming expertly underwater, but are nearly helpless on land. Loons are larger than grebes; they are dagger-billed birds that nest in the Far North and winter mostly on salt water. Some grebes are widespread inland all year, even on small marshy ponds.

Herons and egrets (family Ardeidae) are long-legged, long-necked, spear-billed birds, usually seen standing in the shallows waiting to snatch fish or other aquatic creatures. Unlike cranes (with which they are often confused), herons and egrets fly with their heads hunched back on their shoulders. Cormorants (family Phalacrocoracidae) are dark, long-necked, long-tailed birds with webbed feet and hooked bills. They pursue their prey by swimming, often below the surface. Pelicans (Pelecanidae) are huge birds with odd bills—long and flat, with an expandable pouch to scoop up fish.

Ducks and Rails

Found worldwide and familiar to everyone, the waterfowl family (Anatidae) includes the swans, geese,

178

and ducks. Waterfowl are sociable and usually migrate in flocks. Geese are often seen feeding on land; the sexes are similar. Male ducks—called drakes—are often brightly patterned, while the females are plain. Dabbling ducks feed with only the head and foreparts submerged, while the diving ducks feed underwater.

The rail family (Rallidae) includes the coots, moorhens, and gallinules—ducklike birds that pump their heads as they swim, as well as the chickenlike rails, which hide in marshes and are more often heard than seen.

Shorebirds and Gulls

Known collectively as shorebirds, the plovers (family Charadriidae) and sandpipers (family Scolopacidae) are mostly brown or gray birds, usually found feeding at the water's edge or in fields. Plovers are short-billed birds with plaintive voices. The more diverse sandpipers (including snipes, godwits, curlews, and others) vary from small to very large, and some have very long bills and legs.

Gulls and terns (family Laridae) are long-winged, mostly gray and white water birds. Most gulls are larger than terns, with blunt-tipped bills and omnivorous habits, usually feeding on the ground or on the water. Terns are graceful aerialists with pointed bills and long tails.

Birds of Prey   Nature's cleanup crew, the vultures (family Cathartidae)
can soar effortlessly for hours, searching for the carrion
on which they feed. They have unfeathered heads, heavy
bills, and long, broad wings.
The hawks (family Accipitridae) are hunting birds with
hooked bills, strong talons, and keen eyesight. This
family includes several distinct groups: the broad-
winged, soaring buteos; the bird-catching accipiters; the
graceful kites; the harriers; and the very large eagles.
The Osprey (subfamily Pandioninae) soars and hovers
above the water and dives feetfirst to catch fish in its
sharp, curved talons.
Falcons (family Falconidae) are slim hunting birds with
pointed wings and long tails, adapted for fast flight.
Typical owls (family Strigidae) and barn-owls (family
Tytonidae) hunt by night and roost by day; they share
the traits of upright stance, forward-facing eyes, hooked
bills, strong talons, and very acute hearing and sight.

Various Landbirds   The grouse, quails, pheasants, and turkeys (family
Phasianidae) are plump gamebirds, living on the ground,
with stout bills and strong legs. Their short, rounded
wings carry them on brief bursts of flight to escape
danger. The well-known pigeons and doves (family

180

Columbidae) are round-bodied, small-headed, and short-billed; they are often seen walking on the ground. The long-tailed cuckoos (Cuculidae) feed on large insects and other creatures. Some are secretive, but the roadrunners are easily observed.

Wide gaping mouths, tiny feet, camouflaged brown plumage, and nocturnal habits mark the nighthawks, which belong to the nightjar family (Caprimulgidae). In contrast, the hummingbirds (family Trochilidae) are tiny, long-billed, and energetic, with iridescent plumage. Hummingbirds hover before flowers to take nectar and small insects. They are mostly tropical, and only one species reaches eastern North America. The kingfishers (family Alcedinidae) are small-footed, large-headed, dagger-billed birds that dive from a hovering position or from a perch above the water to catch fish.

Clinging to tree trunks with their toes and propping themselves upright with stiff tail feathers, the woodpeckers (family Picidae) use their long, chisel-like bills to seek insects in and below the bark, and to dig nesting holes in dead trees.

Songbirds

All of the remaining families are classified as perching birds or "songbirds," although some are not particularly good singers. The flycatchers (family Tyrannidae), including the kingbirds and phoebes, are mostly drab in color but have distinctive voices. They usually sally forth from exposed perches to catch insects. Swallows and martins (family Hirundinidae), while foraging in continuous graceful flight rather than from perches, are usually seen in flocks and often near water.

Named for the odd, waxlike tips on certain wing feathers, the waxwings (family Bombycillidae) are slim, crested birds with soft voices. Living mostly on berries and small fruits, they are highly sociable.

The chickadees and titmice (family Paridae) are small, highly active birds that seek insects in trees from the trunks to the outermost twigs. Most have whistled songs and fussing call notes. Often flocking with them in winter, nuthatches (family Sittidae) are small, short-tailed, chisel-billed birds that clamber up, down, or around tree trunks in search of insects.

The crows, jays, and ravens (family Corvidae) are familiar, large birds with heavy bills, omnivorous habits, and often harsh voices. The wrens (family Troglodytidae) are small, hyperactive birds with thin bills and distinctive voices. They often hold their tails up over their backs or flip them about expressively. The mimic thrushes (family Mimidae), including the mockingbirds, catbirds, and thrashers, are slim birds with long tails and thin, pointed bills. They are often seen feeding on the ground, and are famous for their rich, variable songs. The dippers (Cinclidae) are rotund, short-tailed birds that swim and walk underwater in rushing mountain streams.

The family Muscicapidae contains several distinct groups. In North America, the old world warblers (subfamily Sylviinae) are represented mainly by the gnatcatchers and kinglets—tiny, insect-eating birds with thin bills. The thrushes (subfamily Turdinae), include the robins, bluebirds, and spotted thrushes; all are fine songsters that feed heavily on fruits and berries as well as insects.

Native to the Old World, the starling family (Sturnidae) includes many colorful members, but our introduced European Starling is rather plain. It shares family traits of sharp-pointed bill and gregarious habits.

The family Emberizidae is so large and diverse that its many different subfamilies are best considered separately. Small, active, and brightly colored, the wood warblers (subfamily Parulinae) feed mainly on insects; they are highly migratory. Only a few of the tropical tanagers (Thraupinae) reach North America; they are thick-billed, colorful birds of the treetops.

Meadowlarks, orioles, bobolinks, grackles, and cowbirds all belong to the diverse blackbird subfamily (Icterinae). They have sharp-pointed bills and omnivorous diets, and usually wear black or warm colors like yellow or orange. Most species occur in flocks outside the nesting season. The Cardinalinae includes the cardinals, grosbeaks, and buntings. These are small, thick-billed seed-eaters; many species sport bright colors. Similar but usually less colorful are the sparrows and their allies (Emberizinae), including towhees and juncos. Sparrows are most commonly seen feeding on the ground, often in flocks. In a separate family but similar to birds in the last two groups, the cardueline finches (family Fringillidae) are colorful birds with thick, seed-cracking bills. Most are very sociable and have distinctive flight calls, and many are quite erratic in their migrations. The weavers (family Passeridae) are Old World birds, represented here by the introduced House Sparrow, noted for its short, thick bill and noisy, gregarious habits.

On the next page is a list of all the families of birds in this book, together with a page number referring you to the text accounts of the species included.

Water Birds
Loons (20), grebes (22, 34), herons and egrets (18), cormorants (24), pelicans (40).

Ducks and Rails
Ducks, geese, and swans (26, 28, 30, 36, 38), rails (32).

Shorebirds and Gulls
Plovers (48), sandpipers (44, 46, 50), gulls and terns (42).

Birds of Prey
Vultures (154), hawks and eagles (156, 158, 160, 162, 164), falcons (170), Osprey (166), typical owls (172, 174), barn-owls (176).

Various Land Birds
Grouse, quails, and pheasants (148, 152), pigeons and doves (142, 144, 146), cuckoos (150), nightjars (168), hummingbirds (116), kingfishers (70), woodpeckers (122, 124, 126).

Songbirds
Tyrant-flycatchers (56), swallows and martins (58, 60), waxwings (72), chickadees and titmice (94), nuthatches (62), crows and jays (68, 136, 138, 140), wrens (118, 120), mimic-thrushes (52, 54), dippers (110), kinglets (98), thrushes (64, 88), starlings (130), wood warblers (74, 78, 96), tanagers (82), blackbirds (76, 84, 128, 132, 134), cardinals, grosbeaks, and buntings (66, 88), sparrows, towhees, and juncos (90, 92, 102, 104, 106, 108), finches (80, 100, 114), weavers (112).

Glossary

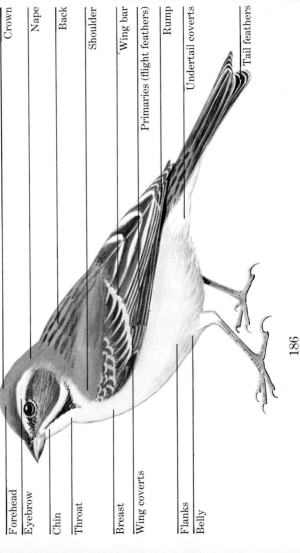

Crown
Nape
Back
Shoulder
Wing bar
Primaries (flight feathers)
Rump
Undertail coverts
Tail feathers

Forehead
Eyebrow
Chin
Throat
Breast
Wing coverts
Flanks
Belly

186

Color phase
One of two or more distinct color types within species, occurring independently of age, sex, or season.

Coverts
The small feathers covering bases of usually larger feathers, providing a smooth, aerodynamic surface.

Crown
The upper surface of the head, between the eyebrows.

Eye-ring
A fleshy or feathered ring around the eye.

Eye stripe
A stripe running horizontally from base of bill through eye.

Mandible
One of the two parts, upper and lower, of a bird's bill.

Mask
An area of contrasting color on front of face and around eyes.

Nape
The back of the head, including the hindneck.

Rump
The lower back, just above the tail.

Speculum
A distinctively colored area on the rear edge of the wing of many ducks.

Underparts
The lower surface of the body, including the chin, throat, breast, belly, sides and flanks, and undertail coverts.

Wing bar
A bar of contrasting color on the upper wing coverts.

Wing lining
A collective term for the coverts of the underwing.

Wing stripe
A lengthwise strip on the upper surface of the extended wing.

Index

Photographers

Roger B. Clapp (133)
Herbert Clark (47, 63, 85, 89, 97, 109, 119, 145)
Harry Darrow (45)
Jack Dermid (29)
Adrian J. Dignan (99)
Larry Ditto (49, 149, 173)
Georges Dremeaux (143)
Kenneth W. Fink (157)
Tim Fitzharris (33, 129, 141)
Jeff Foott (23)
Isidor Jeklin (61, 163, 167, 169)
G.C. Kelley (27, 31, 41, 137, 147, 175, 177)
Wayne Lankinen (21, 25, 39, 103)
Frans Lanting (159, 165)
Tom and Pat Leeson (91)
Thomas W. Martin (107, 113, 131)
Wyman P. Meinzer (55, 151)
Anthony Mercieca (66)
C. Allan Morgan (59)
James F. Parnell (105)
Betty Randall (3, 17, 19, 87, 135)
C. Gable Ray (77, 111)
Laura Riley (153)
Leonard Lee Rue III (139)
C.W. Schwartz (117)
John Shaw (51, 101)

Perry D. Slocum (35, 69)
Arnold Small (43, 57, 83, 93)
Tom Stack & Assoc.
Brian Parker (37)
Alvin E. Staffan (75, 171)
Lynn M. Stone (121)
John Trott (125)
Wardene Weisser (53, 95, 123, 155)
Jack Wilburn (67, 115, 127)
Gary R. Zahm (161)
Leonard Zorn (71, 73, 79)
Jack Zucker (81)

Illustrators

Range maps by Paul Singer
Drawing p. 186 by Lars Svensson

Chanticleer Staff

Publisher: Paul Steiner
Editor-in-Chief: Gudrun Buettner
Executive Editor: Susan Costello
Managing Editor: Jane Opper
Senior Editor: Ann Whitman
Natural Science Editor: John Farrand, Jr.
Associate Editor: David Allen
Assistant Editor: Leslie Marchal
Production: Helga Lose,
Gina Stead
Art Director: Carol Nehring
Art Associate: Ayn Svoboda
Picture Library: Edward Douglas

Design: Massimo Vignelli

The Audubon Society

The National Audubon Society is among the oldest and largest private conservation organizations in the world. With over 515,000 members and more than 500 local chapters across the country, the Society works in behalf of our natural heritage through environmental education and conservation action. It protects wildlife in more than seventy sanctuaries from coast to coast. It also operates outdoor education centers and ecology workshops and publishes the prizewinning AUDUBON magazine, AMERICAN BIRDS magazine, newsletters, films, and other educational materials. For further information regarding membership in the Society, write to the National Audubon Society, 950 Third Avenue, New York, New York 10022.